Teaching mental strategies

number calculations in Years 3 and 4

by Fran Mosley,
Sheila Ebbutt
and Mike Askew

BEAM Education

our thanks to

teachers from Islington, Camden
and Brent for their incisive comment
and continuing support

Published by BEAM Education
© BEAM Education 2001
All rights reserved
ISBN 1 903142 19 9

Designed and typeset by BEAM Education

Printed in England by Beshara Press

British Library Cataloguing-in-Publication
Data

A catalogue record for this publication is
available from the British Library

other books in the series

Teaching mental strategies
number calculations in Years 1 and 2
ISBN 1 903142 18 0

Teaching mental strategies
number calculations in Years 5 and 6
ISBN 1 903142 20 2

Foreword

challenging children to think

The temptation to just 'teach the objectives', when you get to the main part of the mathematics lesson – and you are trying to implement the National Numeracy Strategy – can be very strong. Teachers who adopt this 'transmission' approach may go through the work that has to be covered addressing only very specific, closed questions to children, and then follow this by giving all children a set of similar exercises to work through on their own. This approach, while useful on occasion, leaves little space for children to develop and explain their own ideas and strategies, either to the class or to other children while working in pairs or groups. Children may start to get the message that in mathematics they only have to remember, and not to think – and they will then find it difficult to select strategies to apply in new situations.

This new series of books from BEAM will help teachers both address the objectives and challenge children to think, to come up with strategies and to discuss and share their ideas with others. *Teaching mental strategies* will help teachers anticipate what strategies may be suggested or which ones to introduce if no-one spontaneously comes up with them. The series therefore addresses the true intentions of the Numeracy Strategy: by encouraging genuinely interactive lessons.

I am sure that teachers will find these books enormously helpful, and I wholeheartedly recommend them.

Professor Margaret Brown

School of Education
King's College London

contents

Introduction
how to teach mental strategies?

an alternative approach

We know that numerate pupils have a broad range of mental strategies to draw on when calculating. So mental strategies are learnable, but what is the best way to teach them? One approach is to itemise as many of the strategies as possible, choose an order to teach them in, work through this list and then apply the strategies. This is a bit like the way that manuals for word-processing packages work.

But anyone who has learnt to word process knows that working through a manual can be quite frustrating. The order in which manuals are put together represents the writer's perception of what is needed, rather than the user's, and these do not always coincide. It is no use being taught about font sizes if what you want to learn is how to cut and paste. Similarly, working through a series of mental strategies may not reflect the order in which the learner needs to encounter them.

An alternative approach to teaching strategies and then applying them is to start with a problem-solving situation and see what strategies are needed. This approach means drawing on the cyclical relationship between learning a strategy and using it. Being presented with a problem to solve may make you realise that your existing strategies are not adequate and provoke you into seeking out new ones. As your knowledge of strategies expands, so you can tackle more problems.

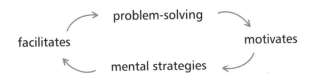

This book provides both problem-solving challenges and ideas for teaching strategies, so that the above cycle can be entered at any point. By starting from a problem-solving situation, pupils can come to appreciate the need for effective strategies and so be motivated to learn them.

how to use this book

Teaching mental strategies presents twenty-four challenges for you to work on with your pupils. Each challenge has been designed to engage children's interest in the mathematics whilst requiring them to use mental strategies appropriate to their age and level of attainment. Each challenge is set out over a double-page spread. The right-hand page provides an outline of the activity and how to set it up in a way that challenges yet engages the class. The left-hand page covers key mental strategies that children are likely to use in solving the challenge and suggestions for how to teach the strategies should the need arise. (More details of how each double-page spread works are given on page 9.)

We recommend that you set the children off on the challenge and give them time to get involved before attending to the strategies that they use.

when to teach the strategies

There are several ways that you can introduce new strategies as the children work on the challenges. Firstly, through your observations of the children as they work on the strategies, you might decide that it is appropriate to intervene and demonstrate an alternative strategy there and then. Even if children are not using the most efficient or effective strategy, it is best to let them carry on using it for a little while, so that they appreciate that there may be a better way to work things out.

Secondly, you may decide to stop everyone from working on the challenge before they reach a conclusion and discuss the various strategies being used. Children can explain their different methods and the class can discuss the merits of each. To get the most out of such a discussion it is better to select in advance who you are going to ask to explain to the class, rather than simply ask for volunteers. That way you can ensure you get a range of strategies and, over time, ensure that different children have the opportunity to report back.

To assist you in keeping track of the different strategies that children use, a photocopiable chart of all the strategies is provided at the back of the book.

Sometimes, you may decide that it is better to let the children work until they resolve the challenge, and then look at the different strategies in the plenary sessions. Once the children have developed a range of strategies you may occasionally decide to go over one or two of the strategies before the children start on the challenge, to remind them about methods.

Using your professional judgement and knowledge of the particular children in your class means you can teach the strategies at different times, so that a balance can be maintained between letting the children use their own methods and moving them on to something more efficient when the time seems right.

learning styles

Many of the mental strategies are needed for more than one challenge. Where this is the case, different styles of teaching the strategies are introduced: some visual, some more oral, some practical. In this way, children's different learning styles can be catered for, and when an individual has difficulty appreciating a strategy presented in one way, you can find a possible alternative. The learning grid at the front of each section will help you see where else the strategies are discussed.

formative assessment

The challenges can also be used as tools for formative assessment. Setting the children off on a challenge and then observing a group of children will provide insight into the thinking of the children. The chart at the back of the book provides a record-keeping device for this.

This page describes the mental strategies that you can teach (or revise) at any point during the challenge.

This page sets out the challenge as a step-by-step classroom activity.

This tells you at a glance which calculation the children will practise.

This is a brief description of the challenge.

This is the equipment you will need.

Each strategy is described in detail, with examples.

Here are the mental strategies that accompany this challenge.

Each challenge is numbered

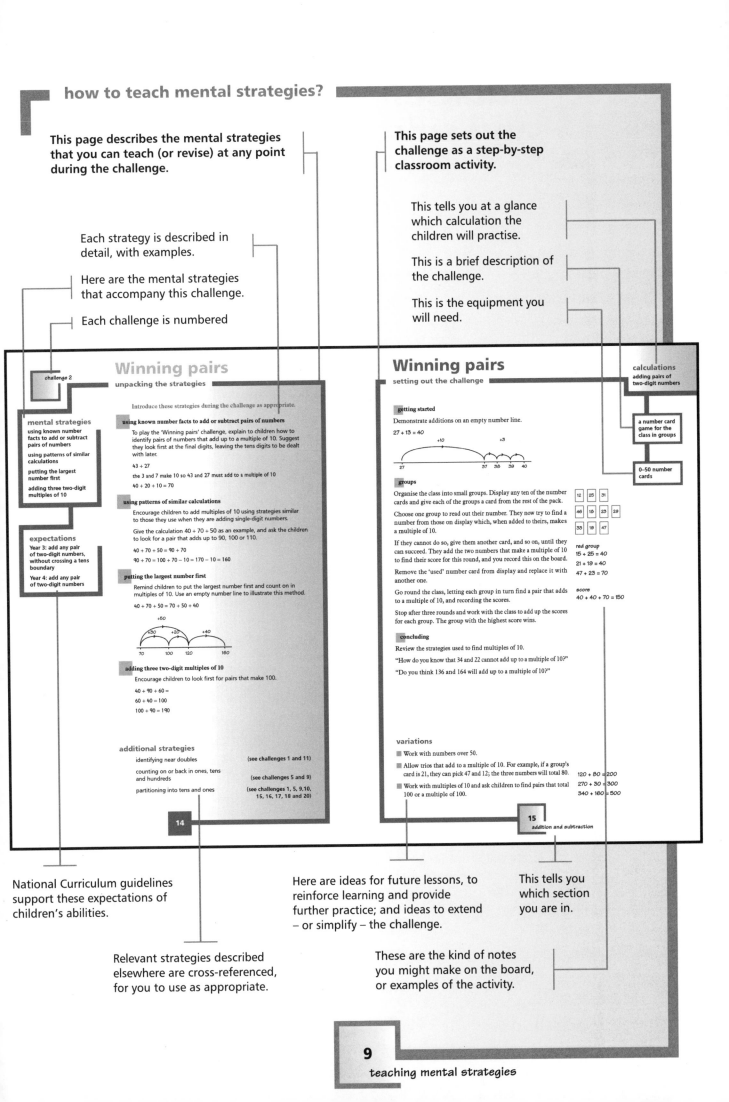

Winning pairs
unpacking the strategies

challenge 2

Introduce these strategies during the challenge as appropriate.

mental strategies

using known number facts to add or subtract pairs of numbers

using patterns of similar calculations

putting the largest number first

adding three two-digit multiples of 10

using known number facts to add or subtract pairs of numbers

To play the 'Winning pairs' challenge, explain to children how to identify pairs of numbers that add up to a multiple of 10. Suggest they look first at the final digits, leaving the tens digits to be dealt with later.

43 + 27

the 3 and 7 make 10 so 43 and 27 must add to a multiple of 10

40 + 20 + 10 = 70

using patterns of similar calculations

Encourage children to add multiples of 10 using strategies similar to those they use when they are adding single-digit numbers.

Give the calculation 40 + 70 + 50 as an example, and ask the children to look for a pair that adds up to 90, 100 or 110.

40 + 70 + 50 = 90 + 70

90 + 70 = 100 + 70 − 10 = 170 − 10 = 160

expectations

Year 3: add any pair of two-digit numbers, without crossing a tens boundary

Year 4: add any pair of two-digit numbers

putting the largest number first

Remind children to put the largest number first and count on in multiples of 10. Use an empty number line to illustrate this method.

40 + 70 + 50 = 70 + 50 + 40

adding three two-digit multiples of 10

Encourage children to look first for pairs that make 100.

40 + 90 + 60 =

60 + 40 = 100

100 + 90 = 190

additional strategies

identifying near doubles — (see challenges 1 and 11)

counting on or back in ones, tens and hundreds — (see challenges 5 and 9)

partitioning into tens and ones — (see challenges 1, 5, 9,10, 15, 16, 17, 18 and 20)

14

Winning pairs
setting out the challenge

calculations
adding pairs of two-digit numbers

a number card game for the class in groups

0–50 number cards

getting started

Demonstrate additions on an empty number line.

27 + 13 = 40

groups

Organise the class into small groups. Display any ten of the number cards and give each of the groups a card from the rest of the pack.

Choose one group to read out their number. They now try to find a number from those on display which, when added to theirs, makes a multiple of 10.

If they cannot do so, give them another card, and so on, until they can succeed. They add the two numbers that make a multiple of 10 to find their score for this round, and you record this on the board.

Remove the 'used' number card from display and replace it with another one.

Go round the class, letting each group in turn find a pair that adds to a multiple of 10, and recording the scores.

Stop after three rounds and work with the class to add up the scores for each group. The group with the highest score wins.

| 12 | 25 | 31 |

| 46 | 18 | 23 | 29 |

| 33 | 19 | 47 |

red group
15 + 25 = 40
21 + 19 = 40
47 + 23 = 70

score
40 + 40 + 70 = 150

concluding

Review the strategies used to find multiples of 10.

"How do you know that 34 and 22 cannot add up to a multiple of 10?"

"Do you think 136 and 164 will add up to a multiple of 10?"

variations

■ Work with numbers over 50.

■ Allow trios that add to a multiple of 10. For example, if a group's card is 21, they can pick 47 and 12; the three numbers will total 80.

■ Work with multiples of 10 and ask children to find pairs that total 100 or a multiple of 100.

120 + 80 = 200
270 + 30 = 300
340 + 160 = 500

15
addition and subtraction

National Curriculum guidelines support these expectations of children's abilities.

Here are ideas for future lessons, to reinforce learning and provide further practice; and ideas to extend – or simplify – the challenge.

This tells you which section you are in.

Relevant strategies described elsewhere are cross-referenced, for you to use as appropriate.

These are the kind of notes you might make on the board, or examples of the activity.

Addition
and subtraction

a set of twelve challenges

the learning grid

calculations

	1 Getting it together	2 Winning pairs	3 The memory game	4 Bull's-eye!	5 Take it away	6 Magic squares	7 Bricklayers	8 Here's one I made earlier	9 Sum difference	10 More or less	11 Doubling up	12 Tens or ones?
adding pairs of two-digit numbers	■	■					■			■		
adding pairs of numbers to 100			■									
adding and subtracting pairs of multiples of 100 and/or 50				■								
subtracting pairs of two-digit numbers					■							
adding several numbers						■						
adding and subtracting a pair of numbers								■				■
adding a multiple of 10 to two- and three-digit numbers									■			
doubling two-digit numbers											■	
finding the difference between pairs of two-digit numbers									■			

strategies

	1	2	3	4	5	6	7	8	9	10	11	12
putting the largest number first		■										
finding a small difference by counting up							■	■	□			■
finding a large difference by counting back from the larger number							■		□			□
counting on or back in ones, tens and hundreds	□	□			■		□		■			□
partitioning into tens and ones	■	□			■				■	■	□	
identifying near doubles	■	□								□	■	
adding and subtracting a near multiple of 10, and adjusting	■		■		■					■		
using the relationship betweeen addition and subtraction			□	□		■	■	■		□		
adding three small numbers by putting the largest number first						■						
adding three small numbers by finding pairs that total 9,10 or 11						■						
adding three two-digit multiples of 10		■										
using known number facts to add or subtract pairs of numbers	□	■	■	■			□	□				
using knowledge of place value to add or subtract pairs of numbers				■		■						
using doubling or halving, starting from known number facts											■	
adding or subtracting a pair of numbers by bridging through 10 or 100						□		■				
using patterns of similar calculations	□	■	■	■								
bridging through a multiple of 10												■

■ this strategy is described in detail here
□ this strategy is referred to as an additional option here

Getting it together

unpacking the strategies

Introduce these strategies during the challenge as appropriate.

mental strategies

identifying near doubles

partitioning into tens and ones

adding and subtracting a near multiple of 10, and adjusting

expectations

Year 3: double any number to at least 20

Year 4: double any whole number from 1 to 50

identifying near doubles

Explain that any pair of horizontal adjacent numbers can be tackled using the strategy 'double one number and adjust'.

$$13 + 14 = 13 + 13 + 1 = 26 + 1 = 27$$

Encourage the children to reverse this process and to recognise that any odd number can be expressed as a sum of near doubles.

$$27 = 26 + 1 = 13 + 13 + 1 = 13 + 14$$

and

$$35 = 34 + 1 = 17 + 17 + 1 = 17 + 18$$

partitioning into tens and ones

Demonstrate how any pair of vertical adjacent numbers can be tackled using the strategy 'add the tens and double the ones'.

$$36 + 46 = 70 + 12 = 82$$

Help the children explore the reverse of this. Any number above 20 can be expressed as the sum of a pair of numbers that differ by 10.

$$82 = 72 + 10 = 36 + 36 + 10 = 36 + 46$$

and

$$36 = 26 + 10 = 13 + 13 + 10 = 13 + 23$$
$$46 = 36 + 10 = 18 + 18 + 10 = 18 + 28$$

adding and subtracting a near multiple of 10, and adjusting

Show the children a strategy for dealing with pairs of numbers that are close to multiples of 10. Encourage them to round the numbers to the nearest multiple of 10 and adjust the answer.

$$39 + 49 = 40 + 50 - 2 = 88$$
$$31 + 41 = 30 + 40 + 2 = 72$$

additional strategies

using known number facts to add or subtract pairs of numbers	(see challenges 2, 3 and 4)
using patterns of similar calculations	(see challenges 2, 3 and 4)
counting on or back in ones, tens and hundreds	(see challenges 5 and 9)

Getting it together

getting started

With a demonstration 100-grid or OHP 100-grid, circle a pair of
adjacent numbers. Numbers directly above and below each other, as
well as numbers side by side, count as adjacent. Ask the children to
add them. Invite children to explain to the class the strategy they
used for adding the numbers.

3	14	15	16	17	18
3	24	25	26	27	28
3	34	35	36	37	38
3	44	45	46	47	48
3	54	55	56	57	58

a pairs activity
developing
confidence in
addition
strategies

pairs

Give pairs of children a 100-grid and three or four cards appropriate
to their level of working.

Give the children five minutes to find pairs of adjacent numbers on
the 100-grid that add up to the numbers on the cards they have been
given. As the children are working, make a note of the strategies they
are using. Decide who you will ask to explain their method to the
class. (Note that it is not possible to find pairs that add up to 21, 41,
61, and so on, simply because of the way the 100-grid is laid out.)

individual
100-grids

20–100
number cards

OHP (optional)

groups

Ask the pairs to join up to form groups of four or six. Suggest they
pool their cards and answers and check each other's work. Ask them
to look for any interesting patterns that emerge.

concluding

Talk with the whole class about how they solved the challenge and
ask selected children to demonstrate.

Ask the children if they can express any generalisations for finding
a pair that total a given number. For example: for an even number,
take off 10, divide the answer by 2 and add 10 to one half; for an
odd number, take off 1, halve it and add 1 to either half.

"What's a quick way of adding any pair of horizontal numbers?"

"Is there a pattern to the answers you get?"

even totals
86 = 48 + 38
68 = 29 + 39
34 = 12 + 22

odd totals
35 = 17 + 18
69 = 34 + 35
43 = 21 + 22

variations

■ Work with a 0–99 grid.

■ Work with diagonal number pairs.

■ Work with trios of adjacent numbers.

Winning pairs

unpacking the strategies

mental strategies

using known number facts to add or subtract pairs of numbers

using patterns of similar calculations

putting the largest number first

adding three two-digit multiples of 10

expectations

Year 3: add any pair of two-digit numbers, without crossing a tens boundary

Year 4: add any pair of two-digit numbers

Introduce these strategies during the challenge as appropriate.

using known number facts to add or subtract pairs of numbers

To play the 'Winning pairs' challenge, explain to children how to identify pairs of numbers that add up to a multiple of 10. Suggest they look first at the final digits, leaving the tens digits to be dealt with later.

43 + 27

the 3 and 7 make 10 so 43 and 27 must add to a multiple of 10

40 + 20 + 10 = 70

using patterns of similar calculations

Encourage children to add multiples of 10 using strategies similar to those they use when they are adding single-digit numbers.

Give the calculation 40 + 70 + 50 as an example, and ask the children to look for a pair that adds up to 90, 100 or 110.

40 + 70 + 50 = 90 + 70

90 + 70 = 100 + 70 − 10 = 170 − 10 = 160

putting the largest number first

Remind children to put the largest number first and count on in multiples of 10. Use an empty number line to illustrate this method.

40 + 70 + 50 = 70 + 50 + 40

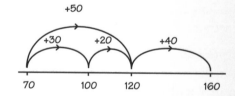

adding three two-digit multiples of 10

Encourage children to look first for pairs that make 100.

40 + 90 + 60 =

60 + 40 = 100

100 + 90 = 190

additional strategies

identifying near doubles   (see challenges 1 and 11)

counting on or back in ones, tens and hundreds   (see challenges 5 and 9)

partitioning into tens and ones   (see challenges 1, 5, 9, 10, 15, 16, 17, 18 and 20)

Winning pairs

getting started

Demonstrate additions on an empty number line.

27 + 13 = 40

groups

Organise the class into small groups. Display any ten of the number cards and give each of the groups a card from the rest of the pack.

Choose one group to read out their number. They now try to find a number from those on display which, when added to theirs, makes a multiple of 10.

If they cannot do so, give them another card, and so on, until they can succeed. They add the two numbers that make a multiple of 10 to find their score for this round, and you record this on the board.

Remove the 'used' number card from display and replace it with another one.

Go round the class, letting each group in turn find a pair that adds to a multiple of 10, and recording the scores.

Stop after three rounds and work with the class to add up the scores for each group. The group with the highest score wins.

concluding

Review the strategies used to find multiples of 10.

"How do you know that 34 and 22 cannot add up to a multiple of 10?"

"Do you think 136 and 164 will add up to a multiple of 10?"

a number card
game for the
class in groups

0–50 number
cards

12	25	31	
46	18	23	29
33	19	47	

red group
15 + 25 = 40

21 + 19 = 40

47 + 23 = 70

score
40 + 40 + 70 = 150

variations

■ Work with numbers over 50.

■ Allow trios that add to a multiple of 10. For example, if a group's card is 21, they can pick 47 and 12; the three numbers will total 80.

■ Work with multiples of 10 and ask children to find pairs that total 100 or a multiple of 100.

120 + 80 = 200

270 + 30 = 300

340 + 160 = 500

The memory game

unpacking the strategies

Introduce these strategies during the challenge as appropriate.

mental strategies

using known number facts to add or subtract pairs of numbers

adding and subtracting a near multiple of 10, and adjusting

using patterns of similar calculations

expectations

Year 3: quickly derive pairs of multiples of 5 that total 100

Year 4: quickly derive pairs of numbers that total 100

using known number facts to add or subtract pairs of numbers

Encourage children to use number facts that they know to figure out other, related facts. There are 51 different pairs of numbers that total 100 (where 90 and 10 give the same total as 10 and 90). Give children practice in recalling known facts to help them derive these pairs easily.

Suggest the children commit 'benchmark' pairs of numbers to memory – for example, the pairs of multiples of 10 that total 100, such as 10, 90 and 20, 80; and as many of the pairs of multiples of 5 as they can, such as 25, 75 and 35, 65.

These can then be used as the basis for 'adding and adjusting the answer'.

$26 + \square = 100$

$25 + 75 = 100$, so it must be 74 that needs to be added

$33 + \square = 100$

$35 + 65 = 100$, so it must be more than 65, it must be 67

adding and subtracting a near multiple of 10, and adjusting

Explain that when a number is close to a multiple of 10 (either above or below it), the quick approach is to work with the multiple of 10 and adjust the answer.

$41 + \square = 100$

$40 + 60 = 100$, so the answer must be 59

$69 + \square = 100$

$70 + 30 = 100$, so the answer must be 31

using patterns of similar calculations

Demonstrate the patterns that can be derived from one known fact by increasing one number by 10 and decreasing the other by 10.

$9 + 91 = 100$

$19 + 81 = 100$

$29 + 71 = 100$
and so on

additional strategies

using the relationship between addition and subtraction

(see challenges 6, 7 and 8)

The memory game
setting out the challenge

a card game
for pairs of
children

blank playing
cards or slips
of paper

getting started

Provide the children with some practice finding the complements
of numbers to 100 (for example, the complement of 26 to 100 is 74).
With Year 3 children, work with the multiples of 5 under 100, and
with Year 4 children work with any two-digit numbers.

Put an array of nine numbers on the board:

Year 3				Year 4		
5	25	40		16	21	36
55	60	85		47	52	78
30	75	10		50	63	97

Ask the children to write down the equivalent arrays of nine numbers,
so that numbers in matching positions total 100.

95	75	60		84	79	64
45	40	15		53	48	22
70	25	90		50	37	3

pairs

In pairs the children produce the cards for 'The memory game'. Each
pair takes ten blank cards and writes down ten numbers under 100.

They mix up the cards, turn each one over in turn and between them
agree what would need to be added to that number to make it up to
100. They write the answer on another blank card.

When the ten answer cards have been made, the children are ready
to play the game. They mix up all twenty cards and leave them
arranged face down on the table. Players take it in turns to turn over
a pair of cards. If the cards total 100 the player keeps that pair. If they
do not total 100, the player turns the cards face down again (and tries
to remember their position on the table). The winner is the player
with more cards when they have all been collected.

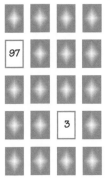

concluding

Go over the strategies that the children used to find complements
to 100.

"How can you know at a glance that 27 and 75 cannot add to 100?"

"How can you use the fact that 25 plus 75 equals 100 to work out
quickly what to add to 26 to make 100?"

variations

- Play the game with pairs of multiples of 100 with a total of 1000.
- Play the game with pairs of multiples of 50 with a total of 1000.

Bull's-eye!

unpacking the strategies

Introduce these strategies during the challenge as appropriate.

mental strategies

using patterns of similar calculations

using known number facts to add or subtract pairs of numbers

using knowledge of place value to add or subtract pairs of numbers

expectations

Year 3: quickly derive pairs of multiples of 100 with a total of 1000

Year 4: quickly derive pairs of multiples of 50 with a total of 1000

using patterns of similar calculations

Work with children on the doubles that they already know. This will help them see patterns in the calculations and empower them to work with larger numbers.

double 3 is 6

so double 300 must be 600

and double 3000 must be 6000

so double 30 is 60

Notice the order here: the language patterns support the understanding that, if double 3 is 6, then double 'three hundred' is 'six hundred'. Although the numbers are smaller, it is not quite so obvious that double 'thirty' must be 'sixty'.

using known number facts to add or subtract pairs of numbers

Provide the children with plenty of practice in rapid recall of bonds to 10, so that when they are adding, say, 700 and 300, the fact that $7 + 3 = 10$ is automatically known.

using knowledge of place value to add or subtract pairs of numbers

Check for understanding of place value and children's confidence in knowing the pairs of numbers that total 10.

Demonstrate patterns on the place value grid (a Gattegno chart) that provide further visual reinforcement.

10 000	20 000	30 000	40 000	50 000	60 000	70 000	80 000	90 000
1 000	2 000	3 000	4 000	5 000	6 000	7 000	8 000	9 000
100	200	300	400	500	600	700	800	900
10	20	30	40	50	60	70	80	90
1	2	3	4	5	6	7	8	9

additional strategies

using the relationship between addition and subtraction

(see challenges 6, 7 and 8)

Bull's-eye!

setting out the challenge

getting started

Work with the children on using known facts for pairs of numbers that total 10 to find totals of multiples of 10 and 100. For example, from $7 + 3 = 10$ the following facts can easily be derived:

$700 + 300 = 1000$

$7000 + 3000 = 10\ 000$

$70 + 30 = 100$

If appropriate, develop patterns from pairs of multiples of 5 that total 100.

$65 + 35 = 100$

$650 + 350 = 1000$

$6500 + 3500 = 10\ 000$

pairs

Children play the 'Bull's-eye!' challenge in pairs. Player One spins the spinner (or rolls the dice). They multiply their score by 100 (Year 3) or 50 (Year 4) and write the answer down followed by the addition sign.

Player Two now spins the spinner, multiplies the number by the same value as Player One and completes the addition. If the answer comes to exactly 1000 (for Year 3) or 500 (for Year 4), then Player Two scores a point. If not, Player One scores a point.

It is now Player Two's turn to spin first.

The winner is the player who has more points after a set number of turns.

concluding

Share children's strategies for quickly spotting the winning pairs of numbers. Encourage children to articulate the relationship between the numbers scored and the final totals: "If the two single digits add up to 10, then the two multiples of 100 total 1000".

"If you know 7 plus 3 equals 10, how does this help you with 700 plus 300?"

"What must I add to 450 to get 1000 exactly?"

variations

■ Player One records a subtraction from 1000 or 500. Round 1: Player One rolls 7 and records $1000 - 700 = \square$. Player Two is trying to score the number that will give the correct answer to the subtraction.

■ Multiply the dice number by 1000 and aim for a total of 10 000.

a game for pairs using place value and addition

1–9 dice or spinners (one per pair)

Year 3

round 1
Player One rolls 7
700 +
Player Two rolls 5
700 + 500 = 1200
Player One scores a point

round 2
Player Two rolls 4
400 +
Player One rolls 6
400 + 600 = 1000
Player Two scores a point

Year 4

round 1
Player One rolls 7
350 +
Player Two rolls 5
350 + 250 = 600
Player One scores a point

round 2
Player Two rolls 4
200 +
Player One rolls 6
200 + 300 = 500
Player Two scores a point

Take it away

unpacking the strategies

mental strategies

counting on or back in
ones, tens and hundreds

partitioning into tens
and ones

adding and subtracting
a near multiple of 10,
and adjusting

expectations

Year 3: add or subtract
any pair of two-digit
numbers, without
crossing a tens boundary

Year 4: add or subtract
any pair of two-digit
numbers

Introduce these strategies during the challenge as appropriate.

counting on or back in ones, tens and hundreds

Check that children know they can use this method – counting on or back in ones, tens and hundreds – for any subtraction. Show them how to support their calculation by drawing an empty number line.

$73 - 41 = \square$

partitioning into tens and ones

Show how to partition the smaller number into tens and ones. Encourage children to use this strategy.

$45 - 32$

$45 - 30 = 15$

$15 - 2 = 13$

so $45 - 32 = 13$

adding and subtracting a near multiple of 10, and adjusting

Explain that, if one number is near to a multiple of 10, a useful strategy is to round it to that multiple, do the subtraction and then adjust the answer.

$33 - 18$

Discuss with the children what needs to be added to adjust the answer if 18 is to be subtracted and 20 is subtracted instead.

$33 - 20 = 13$

so $33 - 18 = 13 + 2 = 15$

Demonstrate the process on a 100 square.

11	12	13	14	15	16	17	18	19	20
21	22	23	24	25	26	27	28	29	30
31	32	33	34	35	36	37	38	39	40
41	42	43	44	45	46	47	48	49	50

additional strategies

using the relationship between
addition and subtraction   **(see challenges 6, 7 and 8)**

Take it away

setting out the challenge

getting started

Show children, or remind them, how they can check the answer to
a subtraction calculation by adding. Demonstrate this on an empty
number line.

"Start with 32 and subtract 4. Where do you end up?"

"Let's check. Add the 4 you subtracted to the 28."

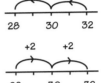

a subtraction
activity for
pairs

Year 3
0–5 number
cards (two
sets per pair)

Year 4
0–9 number
cards (one set
per pair)

pairs

Each pair of children needs a set of 0–9 number cards. (For Year 3
use two sets of 0–5 cards.)

The children shuffle the cards and deal four each. On their own, each
child arranges the four cards to make two two-digit numbers and
writes these out as a subtraction. They then rearrange the cards to
make three more subtraction problems.

| 8 | 2 | | 1 | 5 |

82 – 15

58 – 12

(You may need to remind some children that the larger number
should always come first.)

85 – 12

51 – 28

The children now swap papers and tackle the four problems posed
by their partners. When they have completed this, they swap back
and check each other's work using addition. As the children work,
encourage them to look at the relative size of the numbers and choose
an appropriate strategy. For example, you might count back 12 from
58, but count forward from 28 to 51.

concluding

Discuss the methods children used. Help children to demonstrate
some of the calculations they tackled on the board, writing out the
stages of their thinking, or using an empty number line.

49 – 16

49 – 10 = 39

39 – 6 = 33

49 – 16 = 33

"Which method do you think is the easiest?"

"Which method do you think is the most efficient?"

"Are these the same?"

variations

- Use just three cards and make up problems of the form TO – O.
- Use cards to make up problems of the form HTO – O, HTO – TO
  or THTO – O.
- Use cards to make up addition problems.

Magic squares

unpacking the strategies

Introduce these strategies during the challenge as appropriate.

mental strategies

adding three small numbers by putting the largest number first

adding three small numbers by finding pairs that total 9,10 or 11

using knowledge of place value to add or subtract pairs of numbers

using the relationship between addition and subtraction

expectations

Year 3: add three or four small numbers by putting the largest number first

Year 4: add three or four small numbers, finding pairs totalling to 10

adding three small numbers by putting the largest number first

Point out that, when adding, the order of numbers is irrelevant, so children are free to tackle the numbers in any order they like – for example, by putting the largest number first.

Discourage children from using the strategy 'counting on in ones' with numbers of this size. Show them how to use an empty number line as a bridge to working mentally.

adding three small numbers by finding pairs that total 9, 10 or 11

When children are looking at the numbers in the magic squares, encourage them to look for a pair that adds up to 9, 10 or 11.

$6 + 5 + 4$

$6 + 4 = 10$ and $10 + 5 = 15$

or $6 + 5 = 11$ and $11 + 4 = 15$

and

$13 + 11 + 6$

$13 + 6 = 19$

$19 + 11 = 30$

using knowledge of place value to add or subtract pairs of numbers

When the magic square adds to a number such as 45, encourage children to use their knowledge of place value to do partitioning – adding or subtracting the tens first. For example, they can partition 45 into $40 + 5$, then partition 40 into, say, $15 + 25$. Or they can partition 45 into $42 + 3$, then 42 into $20 + 22$.

using the relationship between addition and subtraction

Discuss how to turn an addition calculation into a subtraction.

$12 + \boxed{} + 4 = 30$

$16 + \boxed{} = 30$

$30 - 16 = \boxed{}$

additional strategies

using known number facts to add or subtract pairs of numbers   **(see challenges 2, 3 and 4)**

adding or subtracting a pair of numbers by bridging through 10 or 100   **(see challenge 8)**

Magic squares
setting out the challenge

getting started

Put a completed 3×3 magic square on the board. Give the children a few minutes to work in pairs and discuss any patterns that they can see in the square. Then, with the whole class, talk about why magic squares are 'magic' and – if no-one knows or has spotted it – point out that each row and column and diagonal adds to the same total (in this case 15).

6	1	8
7	5	3
2	9	4

a number puzzle for pairs

squared paper (optional)

class

Invite individuals to add the rows mentally to check their totals. Talk about the possible ways of doing these additions: choosing the largest number to start with, looking for pairs that make 9 or 10, and so on.

When the rows have been checked, work with the children to check the columns and diagonals.

pairs

Show the class an incomplete magic square. Can they work out what the missing number is? (The magic square shown here is made by adding 5 to each of the numbers of the magic square used at the start of the challenge, but you could add any number as long as you use the same number throughout the square.)

11	6	13
12		8
7	14	9

$11 + 6 + 13 = 30$

The children work in pairs to check each row, column and diagonal and to fill in the missing number (in this case, 10).

Put some other incomplete magic squares on the board, increasing the number of missing numbers. (As long as there are at least four numbers on the square, three of which are in a row, then the other numbers can be figured out.)

12		
	10	
4		8

concluding

Bring the class together to discuss their strategies.

"I want to total 12, 10 and 8. Which pair of numbers would I add first? Why?"

"A magic square has a total of 45. Give me three numbers that might be in a row."

12	2	16
	10	

variations

- Give children a simpler magic square to complete, made by adding 1 to each number of the original.

- Give children a harder magic square to complete (such as one made by multiplying each number by 3 and adding 2).

Bricklayers

unpacking the strategies

Introduce these strategies during the challenge as appropriate.

mental strategies

using the relationship between addition and subtraction

finding a small difference by counting up

finding a large difference by counting back from the larger number

expectations

Year 3: find what must be added to any two-digit number to make the next higher multiple of 10

Year 4: find what must be added to any two-digit number to make 100

using the relationship between addition and subtraction

When children are working on the 'Bricklayers' challenge, they will need to split a larger number such as 50 into two smaller ones. Explain that they can choose a smaller number (say, 26) and from there work out what the other one is. Show the children how problems like these can be written up in various forms, giving them a choice as to how to solve the problem.

26 + ☐ = 50

50 − 26 = ☐

finding a small difference by counting up

Demonstrate that, when the difference between two numbers is small, it makes sense to count up from the smaller number to the larger.

For example, the difference between 50 and 38 is small, so it makes sense to count up from 38.

50 − 38 = ☐

38 + ☐ = 50

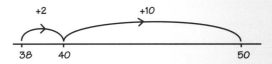

finding a large difference by counting back from the larger number

Point out that, when the difference between two numbers is large, it makes sense to count back from the larger number.

For example, the difference between 50 and 12 is large, so it makes sense to count back 12 from 50.

12 + ☐ = 50

50 − 12 = ☐

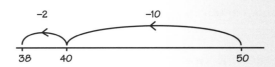

additional strategies

using known number facts to add or subtract pairs of numbers **(see challenges 2, 3 and 4)**

counting on or back in ones, tens and hundreds **(see challenges 5 and 9)**

Bricklayers

setting out the challenge

getting started

Draw a three-course wall on the board or OHP. Invite children to suggest three small numbers and write these in the three bricks at the bottom of the wall.

Show how the wall is built up by adding the two bricks that support the one above and writing the answer in the supported brick.

Repeat this, putting the same three numbers in the bottom row in a different order and asking the children to work out what the top number now is.

Can they find another different arrangement of the three bottom numbers?

Now write 10 in the top, leaving all the others blank, and invite the children to play the 'Bricklayers' challenge with you. They have to find as many different sets of three numbers as they can for the bottom row that will give a top total of 10.

pairs

Now write 50 in the top brick and invite the children to play the 'Bricklayers' challenge in pairs. This time they have to find as many different sets of three numbers as they can for the bottom row that will give a top total of 50.

concluding

Share children's strategies for splitting numbers.

Discuss the fact that the bricks at the bottom don't add up to 50, but, if the value of the bottom middle brick is counted twice, then they do. Try to reach a generalisation that describes this, such as, if you add the three bottom bricks and add the middle bottom brick again, you get the number for the top brick.

"If I change the order of the numbers at the bottom, do I still get 50?"

"I put 2, 4 and 6 in the bottom row of bricks. Can anyone tell me what the top number is without writing it down?"

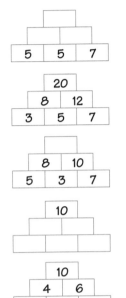

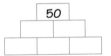

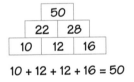

10 + 12 + 12 + 16 = 50

a number puzzle for pairs

squared paper (optional)

variations

■ Put different multiples of 10 in the top brick. Children are free to split the numbers in 'easy' ways.

■ Work with a top total that is a three-digit multiple of 10 on a brick wall with four courses.

Here's one I made earlier

unpacking the strategies

Introduce these strategies during the challenge as appropriate.

mental strategies

using the relationship between addition and subtraction

finding a small difference by counting up

adding or subtracting a pair of numbers by bridging through 10 or 100

expectations

Year 3: subtract any three-digit number from any three-digit number when the difference is small

Year 4: subtract any four-digit number from any four-digit number when the difference is small

using the relationship between addition and subtraction

Give children practice in all the different ways that a related set of additions and subtractions can be expressed. Knowing that $457 - 449 = \square$ can be thought of as $449 + \square = 457$ encourages a counting-on strategy to be used mentally, rather than going immediately to a paper-and-pencil method.

$457 - 449 = \square$

$449 + \square = 457$

$457 - \square = 449$

$\square + 449 = 457$

finding a small difference by counting up

Encourage children to look at the size of the numbers involved and choose a strategy that is appropriate. Explain that counting up from the smaller number to the larger is the most efficient method when the numbers are large and close together. Even when the difference is small, suggest children bridge through a multiple of 10 rather than count on in ones.

$376 - 368$

$368 + 2 = 370$ and $370 + 6 = 376$

so $368 + 8 = 376$

$376 - 368 = 8$

Use an empty number line to model this.

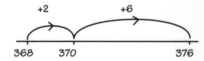

adding or subtracting a pair of numbers by bridging through 10 or 100

Remind children how to add or subtract a single-digit number to or from a two- or three-digit number by using two steps, and bridging through the tens or hundreds.

$68 + 7 =$

$68 + 2 = 70$

$70 + 5 = 75$

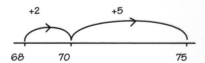

additional strategies

using known number facts to add or subtract pairs of numbers   **(see challenges 2, 3 and 4)**

Here's one I made earlier

setting out the challenge

getting started

Demonstrate this challenge at the board or OHP. Invite a child to come and select four (Year 3) or five (Year 4) number cards. (The challenge is described here for four cards. If you are working with five, you will need to create a four-digit number.)

| 3 | 5 | 6 | 9 |

396

396 + 5

a place value
challenge for
pairs

Ask the child to arrange three of the digits to make a three-digit number. Record this number and complete an addition sentence by adding on the remaining single digit. Ask the class to work out the answer and invite one or two children to explain how they figured it out. If necessary show them how an empty number line can help.

396 + 5 = 401

0–9 number
cards (one set
per pair)

Work together with the children on how this addition can be rearranged as a subtraction calculation.

401 – 396 = 5

demonstration
0–9 number
cards

Repeat this, using the same four cards but arranging them in a different order.

OHP (optional)

pairs

Teaming up in pairs and sharing a set of number cards, the children take four (Year 3) or five (Year 4) each. The challenge starts with the children working individually. Without showing their partner, each child uses their cards to make five different additions of a three-digit number (Year 3) or a four-digit number (Year 4) and a single-digit number. They then figure out the answers to their additions and – on a fresh piece of paper – write these out again as five subtractions.

| 1 | 2 | 5 | 6 |

256 + 1 = 257
265 + 1 = 266
621 + 5 = 626
126 + 5 = 131
152 + 6 = 158

The children swap subtractions with their partner and answer the calculations they are given. They then swap papers back again and check each other's work.

As the class are working on this, observe the different calculation strategies that they are using. Aim to get a range of methods so that children can discuss the merits of each.

257 – 256 =

266 – 265 =

626 – 621 =

131 – 126 =

158 – 152 =

concluding

Invite the children you had identified to explain their methods.

"Which method do you think was the most effective?"

"If you know 432 minus 423 equals 9, what addition calculations do you know?"

variations

■ Take three cards each and make up multiplications of two-digit numbers by single digits and their related divisions.

■ How many different additions of a three-digit number and single digit can be made from four digits, five, six and so on?

Sum difference

unpacking the strategies

Introduce these strategies during the challenge as appropriate.

mental strategies

partitioning into tens and ones

counting on or back in ones, tens and hundreds

expectations

Year 3: add or subtract any single digit to any two-digit number

Year 4: add or subtract any pair of two-digit numbers

partitioning into tens and ones

When children add two numbers, they often partition both the numbers, add the tens and ones separately and then recombine them.

$37 + 23$

$30 + 20 = 50$ and $7 + 3 = 10$

so $37 + 23 = 50 + 10 = 60$

While this works, encourage the children to keep the first number whole and only partition the second.

$37 + 20 = 57$

$57 + 3 = 60$

Show how partitioning only the second number works for subtraction.

$37 - 23$

$37 - 20 = 17$ and $17 - 3 = 14$

so $37 - 23 = 14$

The empty number line provides a visual image of this strategy.

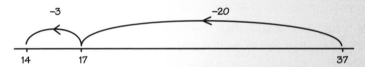

counting on or back in ones, tens and hundreds

Encourage children's fluency in counting forwards and backwards in tens from various starting points.

$0, 10, 20, 30, \ldots 90, 100, 110, \ldots 200$ and back again

$6, 16, 26, 36, \ldots 96, 106, 116, 126$ and back again

$78, 68, 58, 48, 38, 28, 18, 8$

$936, 926, 916, 906, 896, 886, \ldots$

additional strategies

finding a small difference
by counting up   (see challenges 7, 8 and 12)

finding a large difference by counting
back from the larger number   (see challenge 7)

Sum difference

setting out the challenge

adding a multiple
of 10 to two- and
three-digit numbers

finding the
difference between
pairs of two-digit
numbers

getting started

Invite a child to the front of the class and ask them to select two cards
from the 1–9 number cards. Write the two digits on the board, say,
3 and 7.

Ask the child to multiply one of these digits by 10, giving 30 and 7.
Work together with the class on finding the sum and difference
between that multiple of 10 and the single digit.

This gives two new numbers: 37 and 23. Now ask the class to find
the sum and difference of these two numbers.

Repeat this, using the same two original digits, but this time multiplying the other digit by 10.

pairs

Pairs select two numbers from a set of number cards and, working
together, multiply one by 10 and find the sum and difference, and
then the sum and difference of the answers. They swap the numbers
over and repeat.

They then repeat the calculation with a different pair of numbers.
Ask them to write down the multiple of 10, the single digit and the
two final answers.

As the children are working, encourage them to look for patterns in
the answers.

concluding

Discuss some of the different methods that children used for finding the sums and difference.

Work on any generalisations that the children come up with and
encourage them to articulate these clearly:

"The second sum is twice the multiple of 10."

"The second difference is twice the single digit."

"How can you test your statement?"

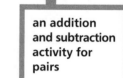

an addition
and subtraction
activity for
pairs

1–9 number
cards

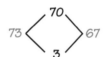

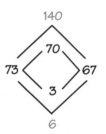

variations

■ A simpler variation is to multiply each digit by 10.

■ A more difficult variation is to select three cards (say, 2, 3, and 4),
multiply one by 10 and put the other two together to create a two-
digit number.

70 + 30 = 100	30 + 24 = 54
70 − 30 = 40	30 − 24 = 6
100 − 40 = 60	54 + 6 = 60
100 + 40 = 140	54 − 6 = 48

More or less

unpacking the strategies

Introduce these strategies during the challenge as appropriate.

mental strategies

partitioning into tens and ones

adding and subtracting a near multiple of 10, and adjusting

expectations

Year 3: add any pair of two-digit numbers, without crossing a tens boundary

Year 4: add any pair of two-digit numbers

partitioning into tens and ones

Tell children that partitioning will help them in the 'More or less' challenge, as it can be used with any pair of two-digit numbers.

Whether the ones total 10 or more or less than 10, the total can always be found by partitioning the smaller number into tens and ones and adding these separately.

$$23 + 41 = 41 + 23 = 41 + 20 + 3 = 61 + 3 = 64$$

Where the ones add to more than 10, encourage the children to bridge through 10.

$$54 + 37 = 54 + 30 + 7 = 84 + 7 = 84 + 6 + 1 = 90 + 1 = 91$$

adding and subtracting a near multiple of 10, and adjusting

Explain that this can be a helpful strategy when dealing with numbers where the ones total more than 10.

For example, $54 + 38$ can be changed to $54 + 40$ by adding an extra 2; then the numbers are added and the extra 2 removed again.

$$54 + 38 = 54 + 40 - 2 = 94 - 2 = 92$$

Demonstrate the method on an empty number line. Point out that it is not vital to draw empty number lines to scale — what matters is that they serve to remind you of where you are in the calculation.

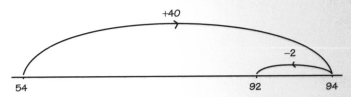

additional strategies

identifying near doubles **(see challenges 1 and 11)**

using the relationship between addition and subtraction **(see challenges 6, 7 and 8)**

More or less
setting out the challenge

calculations
adding pairs of
two-digit numbers

getting started

Draw four boxes on the board with an addition sign between them.

Demonstrate the game on the board, playing against the class. Explain that you are going to roll the dice to generate four digits and you and the class will take turns to put each digit in one of the boxes.

Your aim is to make two numbers which total more than 77: the aim of the class is to achieve a total less than 77.

Roll the dice, read out the number, and invite the class to choose which box to put the digit in. You choose where to put the next one, the class chooses where to put the one after that and you place the last digit.

Discuss with the class how to add the numbers mentally, and write the answer on the board. Agree who has won this round.

		+		

a number challenge the class can play against the teacher

1–6 or 1–9 dice

3	2	+	2	5

total 57;
the class wins

class

Play five more rounds, alternating whether you or a class representative starts the game. As the children become familiar with the game, discuss the choices they make, and help them predict who will win the round.

"I've rolled a 6. If I put it in this box what will happen? Who will win this round?"

pairs

Invite the children to play the game in pairs.

concluding

Divide the class into two and play the game between the two teams.

"I've just rolled a 5. How do I know that I am definitely going to win?"

"I've got 32 as the first number. What's the smallest winning number for the second?"

variations

- Use 0–5 or 0–4 dice or spinners for Year 3, so that adding the ones digits is straightforward and doesn't involve crossing the tens boundary.

- Write out the boxes for four rounds of the game before starting to play. When players roll the dice they may choose any of the boxes to put their number in.

Doubling up

unpacking the strategies

mental strategies

using doubling or halving, starting from known number facts

identifying near doubles

expectations

Year 3: doubling any whole number to 20 and multiples of 5 to 100

Year 4: doubling any whole number to 100

Introduce these strategies during the challenge as appropriate.

using doubling or halving, starting from known number facts

Give children practice in doubling and remind them that they can tackle it in stages. Tell them to do a bit at a time, and to remember after doubling the tens and then doubling the ones to add the separate doubles together.

$17 \times 2 = (10 \times 2) + (7 \times 2)$

$= 20 + 14$

$= 34$

Point out that they can check their answers by carrying out the calculation in a different order.

$10 + 10 = 20$

$20 + 7 = 27$

$27 + 7 = 27 + 3 + 4 = 30 + 4 = 34$

Demonstrate that there will be times when the most efficient strategy is not going to be doubling the tens and ones separately. Explain that, when the number to be doubled is close to a multiple of 10, it may be more efficient to round to the multiple of 10, double and then adjust the answer.

$29 \times 2 = (30 \times 2) - 2 = 60 - 2 = 58$

$41 \times 2 = (40 \times 2) + 2 = 80 + 2 = 82$

identifying near doubles

Show that, from any given double children can derive two 'near doubles'. Explain that they can figure this out by adjusting the answer to the original double.

$25 + 25 = 50$

$24 + 25 = 50 - 1 = 49$

$25 + 26 = 50 + 1 = 51$

additional strategies

partitioning into tens and ones

(see challenges 1, 5, 9, 10, 15, 16, 17, 18 and 20)

Doubling up
setting out the challenge

getting started

Write on the board a two-digit number that the children can easily double, say 20. Discuss the two near doubles that you can get by adjusting double 20, and how the answers relate to each other.

20 + 20 = 40

19 + 20 = 39

20 + 21 = 41

Repeat for two or three other numbers.

a doubling activity for the class in groups

1–100 number cards (optional)

groups

Divide the children into groups of three or four. Each child writes down three numbers (or however many are needed so that each group has a total of twelve numbers) below 50. The group pools their numbers and writes them out in order. If there are any duplicates, suggest they choose a new number; they should end up with a string of twelve.

4   8   12   13   25   32   33   37   40   45   47   48

Call out a number below 100 (you could use a pack of cards and pick one at random) and invite the groups to get as close to this target number as they possibly can. They do this by taking one of their numbers and doubling it, or creating one of the near doubles. For example, you select 48. The group with the example numbers shown above can make 49 by choosing 25 and creating the near double, 24 + 25.

Groups compete against each other for points. Points are awarded when any group can double or create a near double that is close to the target number.

Continue calling out numbers below 100 until one of the groups has won ten points. (Or play for a set amount of time and see which group has scored the most.)

points system
- the target number:
  4 points
- 1–2 more or less:
  2 points
- 3–5 more or less:
  1 point

concluding

Discuss with the class the methods they have used to calculate doubles and near doubles.

"What would be a good set of twelve numbers to have?"

"What two numbers would allow me to get 23 exactly?"

variations

- Children can add two or more of their numbers together before they double or create a near double.

- Children choose numbers between 100 and 200, and you choose numbers between 200 and 400.

Tens or ones?

unpacking the strategies

Introduce these strategies during the challenge as appropriate.

mental strategies

bridging through a multiple of 10

finding a small difference by counting up

expectations

Year 3: multiply single-digit numbers by 10

Year 4: bridge through 100

bridging through a multiple of 10

Encourage children to use the multiples of 10 to 'bounce off' in the calculation. For example, if a child has a running total of 138, and they need to add on 6, counting on in ones is an inefficient strategy. Discuss how 138 is only 2 away from 140.

Build on this work with practice adding single-digit numbers to two-digit numbers.

to do 18 + 5, break down the 5 into 2 and 3

add the 2 to leapfrog onto the 20, then add the remaining 3

$18 + 5 = 18 + 2 + 3 = 23$

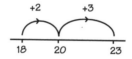

Extend this to the addition of single digits to larger numbers: for example, 48 + 7, 97 + 6 or 136 + 8.

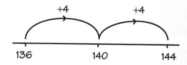

$136 + 8 = 136 + 4 + 4 = 144$

finding a small difference by counting up

This strategy is very likely to crop up in the 'Tens or ones?' challenge: carry out one or more operations which you know will bring you close to the target, then find how much further there is to go by counting up.

$5 + 10 + 10 + 10 = 35$

How much further to go to reach 50?

Show the children how drawing an empty number line when counting up is helpful.

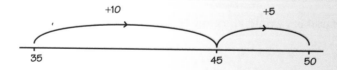

additional strategies

counting on or back in ones, tens and hundreds

(see challenges 5 and 9)

finding a large difference by counting back from the larger number

(see challenge 7)

Tens or ones?

a place value dice game for pairs of children

1–9 dice

OHP (optional)

getting started

Demonstrate the challenge at the board or OHP with you playing against a member of the class.

Ask a child to choose a target number from 100 to 200, and write this on the board.

Take it in turns to roll the dice and announce the number. That player can use the number as it is, or multiply it by 10.

Repeat, with each player keeping a running total of their score. Record as much or as little of this on the board as you see fit.

Play stops when they roll a number that would take them beyond the target number. The score for each player is the difference between their final number and the target. The lowest score wins.

target 129

Player One				Player Two		
dice throw	number	total so far		dice throw	number	total so far
5	50	50		3	30	30
1	10	60		7	70	100
4	40	100		4	4	104
2	20	120		2	20	124
6	6	126		1	1	125
6	bust			7	bust	

final score 129 – 126 = 3   final score 129 – 125 = 4

pairs

Children play several rounds of the game in pairs.

concluding

Play a game against the class. Discuss how, when they are getting close to their target number, they decide what to do.

"The target is 129 and my score so far is 123. Do you think I'll go bust on my next go?"

"If my total were 143, could I have got there in three turns? How?"

variations

■ Aim for a target below 100, using a 1–6 dice. Children can model their additions on a 0–100 number line.

■ Start with a number between 100 and 200. Roll the dice and subtract the number (or multiply it by 10 and subtract that number). Aim to get as close to zero as possible.

Multiplication and division

the learning grid

calculations

	13 Aiming high	14 Aiming low	15 Take the lift	16 Break even	17 Splitting up	18 Laying the table	19 Roll on	20 Hydra	21 Division Bingo	22 Multiplication Bingo	23 Nine eleven	24 Boxed in
multiplying by partitioning	■							■				
deriving doubles of whole numbers and corresponding halves			■	■	■							
multiplying and dividing by 2, 3, 4, 5 and 10		■					■			■		
multiplying and dividing whole numbers by 10 or 100									■			
deriving multiplication facts in the 6 times table and the 8 times table						■						
multiplying by 9 or 11											■	■

strategies

	13 Aiming high	14 Aiming low	15 Take the lift	16 Break even	17 Splitting up	18 Laying the table	19 Roll on	20 Hydra	21 Division Bingo	22 Multiplication Bingo	23 Nine eleven	24 Boxed in
counting on or back in ones, tens and hundreds	□											
doubling or halving two-digit numbers by doubling or halving the tens first			□	■	□		□					
multiplying by 4 by doubling and doubling again						■	■	■		■		
multiplying by 3 by doubling and then adding on the number						□				■		
multiplying by 5 by multiplying by 10 and then halving						■		■		■		
multiplying by 20 by multiplying by 10 and then doubling							■					
finding the 8 times table by doubling the 4 times table						■		■				
finding quarters by halving halves		■		□	■							
using closely related facts to multiply and divide			■		■	□	□		■			
multiplying by 9 by multiplying by 10 and adjusting											■	■
multiplying by 11 by multiplying by 10 and adjusting											■	■
partitioning into tens and ones			■	■	■	■		■		□		
using the relationship between multiplication and addition	■											
using the relationship between multiplication and division		■	□	■	□				□			
multiplying by 10 and 100 by shifting the digits to the left	■						■				□	□
dividing by 10 and 100 by shifting the digits to the right		■							■			
using doubling or halving, starting from known number facts			□									
identifying near doubles								□				

■ this strategy is described in detail here
□ this strategy is referred to as an additional option here

Aiming high

unpacking the strategies

Introduce these strategies during the challenge as appropriate.

mental strategies

multiplying by 10 and 100 by shifting the digits to the left

using the relationship between multiplication and addition

expectations

Year 3: recall multiplication facts for 2, 5 and 10 times tables

Year 4: recall multiplication facts for 2, 3, 4, 5 and 10 times tables

multiplying by 10 and 100 by shifting the digits to the left

Children should know their 10 times table by now, and be able to multiply any digit by 10, simply shifting the digit one place to the left.

Help children recognise that, in the 'Aiming high' challenge, multiplying the starting number by 10 is a 'quick' way of moving towards their target.

4×10

move the 4 one place to the left

40

Use a place value grid to explore the effect of multiplying and dividing by 10 or 100 in terms of moving up and down the rows.

10 000	20 000	30 000	40 000	50 000	60 000	70 000	80 000	90 000
1 000	2 000	3 000	4 000	5 000	6 000	7 000	8 000	9 000
100	200	300	400	500	600	700	800	900
10	20	30	40	50	60	70	80	90
1	2	3	4	5	6	7	8	9

Encourage exploration with a calculator, where children can see that the digits in the display shift to the left when multiplied by 10.

using the relationship between multiplication and addition

Remind children that repeated addition is the same as multiplying. Children may mentally add rather than multiply. When the numbers become large, multiplication becomes more efficient. For example, children may add $10 + 10$ to find 2×10, but use multiplication to find 5×5.

additional strategies

counting on or back in ones, tens and hundreds

(see challenges 5 and 9)

Aiming high
setting out the challenge

a multiplication and addition challenge encouraging lateral thinking

getting started

Write this up on the board:

target 50

2   5   10

+   x

Tell the children that the target is to reach 50 exactly, using the numbers 2, 5 and 10 (as often as necessary) and the operations of addition and multiplication.

Explain that the digits may not be combined to make two-digit numbers – for example, 2 and 5 may not be put together to make 25.

Ask for suggestions how to reach the target of 50. Children are likely to suggest 5×10 as one method. Encourage suggestions of other methods, and collect these on the board. As children explain their methods, help them set the solution out using appropriate notation.

Discuss with the whole class the strategies the children used: how do they know 25×2 is 50? how do they know how many tens they need to add?

target 50

$5 \times 10 = 50$

$(2 \times 10) + (2 \times 10) + 10 = 50$

$(2 \times 5) + (5 \times 5) + (2 \times 5) + 5 = 50$

$(5 \times 5) + (5 \times 5) = 50$

pairs

Children work in pairs on a different target and appropriate numbers, according to the tables they are working on. Tell them to find at least three ways of reaching the target.

target 60

2   3   4

+   x

concluding

Discuss which solutions are neat and quick, and which are quirky and interesting. Emphasise that for most purposes neat and quick solutions are best, but that sometimes – especially in exploratory work such as this – it can be interesting to look at other methods for finding a solution.

"I am using 2, 5 and 10. Will I be able to get to 100 in exactly three calculations?"

"If I calculate 2 times 5 times 5, does it matter which calculation I do first? What about 2 plus 5 times 5?"

variations

- Start at 5. Aim for 50, using 5, 10, and addition.

- Start at 5. Aim for 50, using any numbers and addition.

- Start at zero, 25, –10 or 32. Aim for 95, 100, 150 or 550, using 2, 4, 5, addition and multiplication.

- Start at zero. Aim for 900, using 2, 4, 10, addition and multiplication.

Aiming low

unpacking the strategies

Introduce these strategies during the challenge as appropriate.

mental strategies

dividing by 10 and 100 by shifting the digits to the right

using the relationship between multiplication and division

finding quarters by halving halves

expectations

Year 3: use knowledge of number facts to divide by 2, 5 and 10

Year 4: use closely related facts to carry out division

dividing by 10 and 100 by shifting the digits to the right

Encourage exploration with a calculator to help children become secure about dividing by 10. The digits in the display do very obviously shift to the right, and children can become familiar with the decimal number that results from dividing a number such as 35 by 10.

using the relationship betweeen multiplication and division

Explain that, as long as they know how to multiply by 10 (or 2 or 5 or 4), then dividing by these numbers will not be a problem. Encourage the children to ask themselves the appropriate multiplication question when trying to figure out a division. For example, to divide 45 by 5, the question children need to ask is "What do I need to multiply 5 by to get 45?".

$45 \div 5 =$

$5 \times \boxed{} = 45$

$5 \times 9 = 45$

so $45 \div 5 = 9$

By understanding this relationship, the children can also use their knowledge of doubles to find halves.

$400 \div 2$

$2 \times \boxed{} = 400$

double 200 is 400 so half of 400 must be 200

$400 \div 2 = 200$

finding quarters by halving halves

Encourage the children doing the 'Aiming low' challenge to make use of all the numbers they are allowed (2, 4, 5 and 10). This will give them practice in dividing by 4. Revise the idea that dividing by 4 is the same as finding a quarter, which they can do by halving and then halving again.

$80 \div 4$

half of 80 is 40 and half of 40 is 20

so a quarter of 80 is 20

$80 \div 4 = 20$

additional strategies

using patterns of similar calculations   (see challenges 2, 3 and 4)

Aiming low

getting started

Write the number 500 on the board.

Explain that you are starting with 500 and you want to reduce it
to zero. Ask for ways to get from 500 to zero using the numbers 2, 4,
5 and 10 (as often as necessary) and the operations of subtraction and
division.

Work with the class to find a solution using a mixture of division
and subtraction, and take the opportunity to revise division by 10.

You should quickly establish that 500 ÷ 10 is one method. Encourage
suggestions of other methods, and collect these on the board.

pairs

Set children a similar task to carry out alone or with a partner. Choose
numbers appropriate to the tables they should be working on. Ask
them to find at least three ways of reaching the target.

concluding

Collect several different solutions for the same starting number.
Encourage and support the children both to explain their method
clearly and to use appropriate notation to record it.

Discuss with the whole class the strategies children used.

"How do you work out 100 divided by 4?"

"How do you know how many tens you need to subtract?"

start at 500

target 0

2   4   5   10

$-$ $\div$

$500 \div 10 = 50$

$50 \div 5 = 10$

$10 - 10 = 0$

$500 \div 5 = 100$

$100 \div 4 = 25$

$25 - 5 = 20$

$20 - 10 - 10 = 0$

start at 450

target 0

2   4   5   10

$-$ $\div$

a subtraction
and division
challenge
encouraging
lateral thinking

variations

- Start at 50 and aim for zero, using 5, 10, and subtraction.

- Start at 100 and aim for zero, using any numbers and subtraction.

- Start at 750, 1000 or 1200. Aim for zero (or 5, −5, 0·5, 0·1) using
  2, 5, 20, subtraction and division.

- Make up a similar problem for a friend to tackle.

Take the lift

unpacking the strategies

Introduce these strategies during the challenge as appropriate.

mental strategies

partitioning into tens and ones

using closely related facts to multiply and divide

partitioning into tens and ones

Partitioning is an important strategy for all the two-digit doublings children carry out in this 'Take the lift' challenge — unless they are doubles children know by rapid recall.

Remind children who lack confidence that a multiplication problem can be carried out in stages. Tell them to tackle a bit at a time, to remember to add the results together and to check their answers.

$24 \times 2 = (20 \times 2) + (4 \times 2)$

$= 40 + 8$

$= 48$

Show how the calculation can be checked using a different method.

$24 \times 2 = (25 - 1) \times 2$

$= 25 \times 2 - 2$

$= 50 - 2$

$= 48$

expectations

Year 3: double any number to at least 20

Year 4: double any number to at least 50

using closely related facts to multiply and divide

Remind children that they can double any number below 100. To do this, all they need is to know the 2 times table, and to understand how to multiply a multiple of 10 such as 30 by a single digit.

9	9×2
18	10×2 added to 8×2 or 20×2 subtract 2×2
36	3×2 is 6 so 30×2 is 60; add on 6×2
72	7×2 is 14 so 70×2 is 140; add on 2×2

additional strategies

using doubling or halving, starting
from known number facts   **(see challenge 11)**

using the relationship between
multiplication and division   **(see challenges 14 and 16)**

doubling or halving two-digit numbers
by doubling or halving the tens first   **(see challenge 16)**

Take the lift
setting out the challenge

getting started

Write up a single digit, such as 4 or 5, and invite the class to help
you double it repeatedly until you get an answer that is beyond 50
(Year 3) or 100 (Year 4).

Talk about how you can do each doubling and elicit the knowledge
children already have that they can make use of. For example, the
children will know the 2 times table and the strategy of splitting
a two-digit number into tens and ones and doubling or multiplying
each part separately.

4
8
16
32
64
(128)

a class activity
on doubling
leading to an
investigation
in pairs

pairs

The children now work in pairs. Each child chooses a single digit
and – without showing their partner their working – repeatedly
doubles the number they have chosen, until they get an answer that
is greater than 50 or 100. They swap the final answer with their partner (still keeping the rest of their calculation a secret): the challenge
is to work out what their partner's starting number was.

concluding

Clarify any confusion about the process of doubling. Record on the
board the doublings for each digit in turn, including 1. Talk about
the fact that doubling zero simply gives zero.

$0 \times 2 = 0$

Discuss the fact that some of the repeated doublings copy each
other. For example, doubling 4 repeatedly is the same as doubling
2 repeatedly, with one fewer number.

Establish that 3 and 6 are the digits which get nearest to 100 when
doubled (the sequence goes 3 6 12 24 48 96).

"What did I double to get 72?"

"I doubled a number and got 83 as my answer. How do I know I've
made a slip-up?"

start with 4	start with 2
	2
4	4
8	8
16	16
32	32
64	64
128	128

variations

■ Which starting digit gets closest to 1000 when it is doubled
repeatedly? Is it the same digit as the one that gets closest to 100?

■ Which starting digit gets closest to 800? To 600? Guess, then check.

Break even

unpacking the strategies

Introduce these strategies during the challenge as appropriate.

mental strategies

doubling or halving two-digit numbers by doubling or halving the tens first

partitioning into tens and ones

using the relationship between multiplication and division

expectations

Year 3: halve any even number to at least 40

Year 4: halve any even number to at least 100

doubling or halving two-digit numbers by doubling or halving the tens first

Where the tens digit is even, tell the children they can deal with the tens and ones completely separately: to halve 46, halve each digit in turn, to get 23.

Clearly, this works with even digits, but is more complicated with odd digits. So 46 is easy to deal with: to halve 40 + 6, you get 20 + 3. Halving 74 involves more complex mental calculations (see below).

partitioning into tens and ones

Numbers with an odd tens digit are slightly more complicated and will involve some kind of partitioning. Children will have different ideas on how to tackle this. Explore the variety of methods with them.

halve 38 by halving 30 then 8

halve 38 by halving 20 and then halving 18

halve 38 by halving 40 and adjusting the answer

using the relationship between multiplication and division

Encourage children to use the reciprocal relationship between doubling and halving – for example, by doubling teens numbers such as 16, 17 and 18 until they find the one that makes 36.

Remind children that, as long as they know the doubles of numbers up to 10, then halving numbers below 20 should be no problem.

$9 \times 2 = 18$

so half of 18 is 9

additional strategies

finding quarters by halving halves

(see challenges 14 and 17)

Break even

calculations

deriving doubles
of whole numbers
and corresponding
halves

getting started

Spend a few minutes practising halving two-digit numbers.

class

Write on the board a two-digit number with an even tens digit. Start with a number below 50 for Year 3 and below 100 for Year 4.

46

a class activity
involving a
generalisation

Tell the class you are going to play the 'Break even' challenge with this number, using the rule: "If the number is even, halve it; if it is odd, add one and then halve it". Check the children are confident about how to identify odd and even numbers.

Ask individual children to help you apply this rule to the number on the board, and to the chain of numbers created, until you reach the number 1.

Start another chain, if you think the children need to have the rules reinforced.

pairs

Challenge the children to find which starting number below either 50 or 100 produces the longest chain. How do the children know this is the longest chain? (The answers are 33 and 65.)

concluding

Invite the children to share the different chains that they found.

"The number 46 ends in 6, 3, 2, 1. Does the chain for any other number end in 6, 3, 2, 1?"

"What's special about the chain produced by 32?"

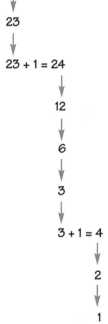

46
↓
23
↓
23 + 1 = 24
↓
12
↓
6
↓
3
↓
3 + 1 = 4
↓
2
↓
1

variations

■ Try numbers over 100 or even over 1000. Does the chain always end in 1? How can they be certain?

Splitting up
unpacking the strategies

mental strategies

partitioning into tens
and ones

finding quarters by
halving halves

using closely related
facts to multiply
and divide

expectations

**Year 3: double any
number to 20 and find
corresponding halves**

**Year 4: double any
number to 50 and find
corresponding halves**

Introduce these strategies during the challenge as appropriate.

partitioning into tens and ones

Tell the children they will have no problems halving a number if they
halve the tens and ones separately.

44

half of 40 is 20 and half of 4 is 2

$20 + 2 = 22$; so half of 44 is 22

This is straightforward if there is an even number of tens. Where
there is an odd number of tens, there is a choice of strategies.

34

half of 30 is 15 and half of 4 is 2

$15 + 2 = 17$; so half of 34 is 17

An alternative strategy is:

34 is $20 + 14$

half of 20 is 10 and half of 14 is 7

$10 + 7 = 17$; so half of 34 is 17

finding quarters by halving halves

Demonstrate to the children that, to find a quarter of a number, they
can halve the number and then see if it is possible to halve it again.

36

half of 36 is 18

half of 18 is 9; so a quarter of 36 is 9

34

half of 34 is 17

I cannot halve 17

using closely related facts to multiply and divide

Point out another way of finding a quarter, and that is by using their
knowledge of multiplication facts to find out a division fact.

to find a quarter of 28

I know that 7×4 is 28

so 7 is a quarter of 28

additional strategies

doubling or halving two-digit numbers
by doubling or halving the tens first   (see challenge 16)

using the relationship
between multiplication and division   (see challenges 14 and 16)

Splitting up
setting out the challenge

an activity
for children in
pairs

Year 3

1–25 number
cards (one set
per pair)

Year 4

1–50 number
cards (one set
per pair)

getting started

Spend a few minutes practising halving and quartering two-digit
numbers. Choose two-digit numbers that are multiples of 4 as this
will avoid answers involving fractions or decimal numbers.

40

$\frac{1}{2}$ *of 40 is 20*

$\frac{1}{4}$ *of 40 is* $\frac{1}{2}$ *of 20*

which is 10

pairs

Children work in pairs to play the 'Splitting up' challenge. Each pair
needs a set of 1–25 (or 1–50) number cards. They mix up the cards
and place them in a pile face down between them. Each player writes
down the digits 1 to 9 (or 1 to 19).

They take it in turns to turn over the top card in the pile. Their aim
is to see if they can find a half or a quarter of the number revealed.
They are aiming to strike out one (only one on each turn) of the digits
they have written down on their piece of paper.

For example, player 1 turns over 28. Player One can quarter this and
strike out 7.

Player Two turns over 16. Player Two can choose between halving
this and striking out 8 or quartering it and striking out 4.

Player One turns over 13 and cannot strike anything out.

The first between them to strike out five numbers has won the
challenge. If time allows, children can play 'Splitting up' three times.
The best out of three wins.

concluding

Using a few sample numbers, review the different strategies the
children have used.

"Tell me some numbers where you can find a quarter. What do you
notice about those numbers?"

"I quartered a number and got 6. What was my number?"

variations

▪ Invent a similar game involving eighths.

▪ Children take it in turns to select a card without showing their
   partner the number. If they can, they halve or quarter the number
   and tell their partner the answer. Can their partner work out the
   original number?

Laying the table

unpacking the strategies

Introduce these strategies during the challenge as appropriate.

mental strategies

multiplying by 4 by doubling and doubling again

finding the 8 times table by doubling the 4 times table

multiplying by 5 by multiplying by 10 and then halving

partitioning into tens and ones

expectations

Year 3: use knowledge of number facts and place value to multiply by 2, 5, 10

Year 4: use closely related facts to carry out multiplications

multiplying by 4 by doubling and doubling again

Encourage children to become familiar with doubling as a method, and with repeated doubling.

If children seem uncertain about doubling as a strategy, do a simple exercise. Ask two children to stand up, each holding up six fingers. Establish with the class that 6×2 is 12 and write this up. Now say you want to double the number of children, and invite another two children to join them, but to stand a little apart from the first pair. Establish that the number of children is doubled from 2 to 4, and so is the number of fingers, from 12 to 24. On the board, write:

$6 \times 2 = 12$

2 lots of $6 \times 2 = 2$ lots of 12

$6 \times 4 = 24$

finding the 8 times table by doubling the 4 times table

Demonstrate how doubling can be a useful strategy by asking a child to multiply a number such as 6 by 8. Show how, if they use repeated doubling, this will help them get around the fact that they are not yet confident with either the 6 times table or the 8 times table.

$6 \times 2 = 12$

$6 \times 4 = 24$ (by doubling the previous result)

$6 \times 8 = 48$ (by doubling the previous result)

multiplying by 5 by multiplying by 10 and then halving

Explain to children that they can find 6×5 by at least two methods. The first involves working out 6×4 (by doubling) and then just adding another 6. The second method is to multiply 6 by 10 and then halve it. Children can use the other method to check their work.

$6 \times 10 = 60$

as 5 is half of 10, then 6×5 is half of 60,

so 6×5 is 30

partitioning into tens and ones

Remind children they can multiply numbers in stages. This means they can use partitioning. For example, they can create the 13 times table by partitioning 13 into 10 and 3.

$13 \times 7 = (10 \times 7) + (3 \times 7) = 70 + 21 = 91$

additional strategies

using closely related facts
to multiply and divide **(see challenges 15, 17 and 21)**

multiplying by 3 by doubling
and then adding the number **(see challenge 22)**

Laying the table

setting out the challenge

getting started

Tell the children that they can work out a multiplication table they don't know just by using facts that they already know.

Write on the board:

$6 \times 1 = 6$

Invite the class to suggest some other facts about the 6 times table that they can work out from this.

If necessary prompt them to offer $6 \times 10 = 60$ (from the 10 times table) and $6 \times 2 = 12$ (from the 2 times table).

Write these in, leaving space for the missing facts.

Now ask the class to suggest other facts that can be worked out from the new facts. They might suggest that 6×9 is 6 less than 6×10 or 6×4 is double 6×2, and then that 6×8 is 6×4 doubled again. Again, write these in, leaving space for the missing facts.

Continue eliciting facts from the 6 times table until it is complete.

Now rub out a few of the answers and invite suggestions as to how to work out the missing numbers.

pairs

Ask the children to create a multiplication table for 7, 8, 9, 15 or 25.

When they have finished, invite the children to create a table using their own choice of number.

concluding

Invite some pairs to demonstrate to the rest of the class how they created their tables. Review the different strategies used.

"10 times 7 is 70. How do I know what 5 times 7 is?"

"Can anyone explain why 10 multiplied by 9 has the same answer as 9 multiplied by 10?"

$6 \times 1 = 6$

$6 \times 2 = 12$

$6 \times 10 = 60$

$6 \times 1 = 6$

$6 \times 2 = 12$

$6 \times 3 =$

$6 \times 4 = 24$

$6 \times 5 = 30$

$6 \times 6 =$

$6 \times 7 = 42$

$6 \times 8 = 48$

$6 \times 9 =$

$6 \times 10 = 60$

variations

- Ask the children to derive the facts in the 8 times table by doubling the facts in the 4 times table.

- Invite the children to check the facts of the 6 times table that they all worked out together by adding the results of the 2 times table and the 4 times table, or the 1 times table and the 5 times table; or by doubling the facts in the 3 times table.

Roll on

unpacking the strategies

mental strategies

multiplying by 4 by doubling and doubling again

multiplying by 20 by multiplying by 10 and then doubling

multiplying by 10 and 100 by shifting the digits to the left

expectations

Year 3: multiply a single digit by 2, 100 and 20

Year 4: multiply a teen number by 2, 100 and 20

Introduce these strategies during the challenge as appropriate.

multiplying by 4 by doubling and doubling again

Use diagrams to demonstrate to the children how multiplying by 4 can be done by multiplying a number by 2 and then multiplying it by 2 again.

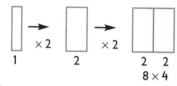

multiplying by 20 by multiplying by 10 and then doubling

Use an overhead calculator to demonstrate how the result of multiplying by 10 and then multiplying by 2 (or doubling) gives the same result as multiplying by 20.

$14 \times 10 = 140$

$140 \times 2 = 280$ and $14 \times 20 = 280$

multiplying by 10 and 100 by shifting the digits to the left

Use a place value board to help the children visualise the effect of multiplying by 10 and then multiplying by 10 again.

Multiplying by 10 moves the digits one place to the left:

Th	H	T	O
		1	5

Th	H	T	O
	1	5	0

$15 \times 10 = 150$

Multiplying by 10 again moves the digits another place to the left:

Th	H	T	O
	1	5	0

Th	H	T	O
1	5	0	0

$150 \times 10 = 1500$

Multiplying by 100 moves the digits two places to the left:

Th	H	T	O
		1	5

Th	H	T	O
1	5	0	0

$15 \times 100 = 1500$

additional strategies

using closely related facts
to multiply and divide   **(see challenges 15, 17 and 21)**

doubling or halving two-digit numbers
by doubling or halving the tens first   **(see challenge 16)**

Roll on

setting out the challenge

getting started

Practise rapid mental multiplication by 10 and by 2.

Then demonstrate the rules of the 'Roll on' challenge by playing
against one of the class (either someone who volunteers or someone
you select).

Invite children to decide at the outset whether they are going to play
for a greater or lesser total.

In the first round Player One rolls the number dice and writes down
the number. Player Two then rolls the operation dice and records the
operation against the number. Player One rolls the operation dice
again, records a second operation and then completes the calculation.

In the second round Player Two rolls the number dice and they play
the game through again.

After six rounds each player totals their three scores. The winner is
the player with the greater or lesser total – depending on what they
decided at the beginning of the game.

pairs

Children play the 'Roll on' challenge in pairs.

concluding

Discuss with the class what each pair of multiplications is equivalent to. There are three possible outcomes: ×4 (×2 ×2), ×20 (×10 ×2)
and ×100 (×10 ×10).

"Do I get a different final product if I multiply by 10 and then by 2
or if I multiply by 2 and then by 10?"

"Why not?"

Finish with some mental calculations involving ×4, ×20, ×100.

round 1

Player One

6

Player Two

6 x 10

Player One

6 x 10 x 2

6 x 10 x 2 = 120

round 2

Player Two

7

Player One

7 x 10

Player Two

7 x 10 x 10

7 x 10 x 10 = 700

a multiplication
game for pairs

Year 3
1–9 dice or
spinner per
pair

dice or spinner
per pair,
marked ×10,
×10, ×10, ×2,
×2, ×2

Year 4
1–20 dice or
spinner per
pair

dice or spinner
per pair,
marked ×10,
×10, ×10, ×2,
×2, ×2

variations

▒ Use the same activity, with different dice, to show how ×6 is the
same as ×2 ×3; ×8 is the same as ×2 ×4; and ×9 is the same as ×3 ×3.

▒ Use a dice marked ×20, ×20, ×20, ×4, ×4, ×4.

Hydra
unpacking the strategies

Introduce these strategies during the challenge as appropriate.

mental strategies

partitioning into tens and ones

multiplying by 4 by doubling and doubling again

finding the 8 times table by doubling the 4 times table

multiplying by 5 by multiplying by 10 and then halving

expectations

Year 3: use knowledge of number facts to multiply by 2, 5

Year 4: use knowledge of number facts to multiply by 2, 3, 4, 5

partitioning into tens and ones

Explain that a multiplication problem can be carried out in stages. Show how to separate the calculation into different bits and remind the children to add the results together.

$$25 \times 3 = (20 \times 3) + (5 \times 3)$$
$$= 60 + 15$$
$$= 75$$

multiplying by 4 by doubling and doubling again

In the second part of 'Hydra', children use the digit 4. Remind them that to multiply by 4, you can double, then double again. Write down part of the 2 times table and the 4 times table, and look at how the answers in the second are all double those in the first.

$2 \times 3 = 6$	$4 \times 3 = 12$
$2 \times 4 = 8$	$4 \times 4 = 16$
$2 \times 5 = 10$	$4 \times 5 = 20$

finding the 8 times table by doubling the 4 times table

Explain that to multiply by 8, all you have to do is double the 4 times table. Write down part of the 4 times table and the 8 times table, and look at how the answers in the second are all double those in the first.

$4 \times 1 = 4$	$8 \times 1 = 8$
$4 \times 2 = 8$	$8 \times 2 = 16$
$4 \times 3 = 12$	$8 \times 3 = 24$

multiplying by 5 by multiplying by 10 and then halving

Show how multiplying by 5 is also easy. If children can multiply by 10, and can halve numbers, multiplying by 5 is straightforward. Write down part of the 5 and the 10 times tables and show how the answers in the one are all half of those in the other.

$5 \times 3 = 15$	$10 \times 3 = 30$
$5 \times 4 = 20$	$10 \times 4 = 40$
$5 \times 5 = 25$	$10 \times 5 = 50$

additional strategies

identifying near doubles

(see challenges 1 and 11)

Hydra

getting started

On the board draw two boxes together, a multiplication sign and one box on its own. Write the digits 2, 3 and 5 alongside.

☐☐ × ☐   2   3   5

Now tell the class you are going to arrange these three digits in the boxes to make a two-digit and a one-digit number. The aim is to find out which combination gives the greatest product and which gives the least.

First, establish all the different possible arrangements of the digits. Ask individual children to suggest new ones until you have all six arrangements. Sort these into an order that helps children see whether there are any missing. Discuss how you can be sure that this is a complete set.

23×5
25×3
32×5
35×2
52×3
53×2

pairs

Children work in pairs to carry out the six multiplications. When they have finished, they put the answers in order, to see which gives the greatest product and which gives the least.

70
75
106
115
156
160

class

Discuss with the children the various multiplication methods they have used.

pairs

Now change the 5 to 4 and ask the children to work in pairs, find all the possible products and arrange these in order from the greatest to the least.

concluding

Again, go over the children's methods and discuss whether they have tried a new method they have not used before.

"What's the best way of splitting up 23 to multiply 23 by 4?"

"How would you work out 23 times 8?"

variations

- Use digits 3, 4 and 5; or use digits 3, 5 and 6 and remind them how to derive the facts in the 6 times table (see challenge 18).

- Use digits 2, 2 and 9 or 2, 3 and 9 or 1, 2 and 9. Discuss ways of multiplying by 9, 19 or 29 such as multiplying by 10 (or 20 or 30) and adjusting the answer.

Division Bingo

unpacking the strategies

Introduce these strategies during the challenge as appropriate.

mental strategies

dividing by 10 and 100 by shifting the digits to the right

using closely related facts to multiply and divide

expectations

Year 3: use knowledge of number facts and place value to multiply and divide by 10 and 100

Year 4: use known number facts and place value to multiply and divide by 10 and 100

dividing by 10 and 100 by shifting the digits to the right

Use a place value grid to explore the effect of dividing and multiplying by 10 or 100 in terms of moving up and down the rows.

10 000	20 000	30 000	40 000	50 000	60 000	70 000	80 000	90 000
1 000	2 000	3 000	4 000	5 000	6 000	7 000	8 000	9 000
100	200	300	400	500	600	700	800	900
10	20	30	40	50	60	70	80	90
1	2	3	4	5	6	7	8	9

Encourage the children to use a calculator to see how the digits in the display shift to the right when divided by 10. Show children the decimal numbers that result from dividing numbers such as 35 or 365 by 10 or 100.

using closely related facts to multiply and divide

Practise other divisions and multiplications, all based on multiplying or dividing by 10 or 100.

to divide by 10	shift the digits one place to the right
to divide by 100	shift the digits two places to the right
to divide by 20	divide by 10 then halve the answer
to divide by 5	divide by 10 and double the answer
to multiply by 10	shift the digits one place to the left
to multiply by 100	shift the digits two places to the left
to multiply by 20	multiply by 10 then double the answer
to multiply by 5	multiply by 10 and halve the answer

additional strategies

using the relationship between multiplication and division

(see challenges 14 and 16)

Division Bingo

setting out the challenge

getting started

Write these two columns of numbers up on the board.

Ask the children to draw a 2 × 2 grid (their 'Bingo card') and to choose four numbers from the right-hand column to write on their card, one number per box. Now rub out the right-hand column.

30	3
40	4
300	30
400	40
370	37
450	45
610	61
720	72

class

Call out, one by one, numbers from the left-hand column. Each time, ask for a volunteer to divide the number by 10 (vary this by asking for 'a tenth of') and announce the answer.

Any child who has that number on their grid should cross it out.

The game is over when one child has crossed out all four numbers (or when everybody has crossed out at least one number).

a Bingo game for the whole class, then groups

slips of paper – eight for each child

groups

Divide the class into groups of three or four children. Individually, each child writes four different numbers under 50 on separate slips of paper. On four more slips they write the results of multiplying each of their numbers by 10.

The group pools the slips with the original numbers on, mixes them up and then shares them out so that everyone has four new numbers. They then separately pool all the answer slips, arranging these face down on the table.

Players take it in turns to choose one of the answer slips and read out the number. Whoever has the result of dividing that number by 10 can keep the answer and their number as a 'trick'. The winner is the first player to get four 'tricks'.

concluding

Discuss with the children how dividing by 10 relates to multiplying by 10 and how this can be extended into multiplying and dividing by 100.

"What happens to the digits in a number when you divide by 10?"

"What happens if you multiply this number by 100?"

variations

- Play Division Bingo where multiples of 100 and 1000 are divided by 10 (or by 100).

- Play Division Bingo where the numbers are divided by 20 – you can divide by 10 and halve the answer. Or divide by 5 – you can divide by 10 and double the answer.

Multiplication Bingo
unpacking the strategies

Introduce these strategies during the challenge as appropriate.

mental strategies

multiplying by 4 by doubling and doubling again

multiplying by 5 by multiplying by 10 and then halving

multiplying by 3 by doubling, and then adding on the number

expectations

Year 3: use knowledge of number facts and place value to multiply by 2, 5, 10

Year 4: use knowledge of number facts and place value to multiply by 2, 3, 4, 5, 10

multiplying by 4 by doubling and doubling again

Go over various alternative ways of tackling one of the multiplications in this Bingo challenge with the children.

For example, 4×25 could be done by partitioning into tens and ones:

$4 \times 25 = (4 \times 20) + (4 \times 5)$

or by the method 'double and double again':

$4 \times 25 = 2 \times 2 \times 25$

$= 2 \times 50$

$= 100$

multiplying by 5 by multiplying by 10 and then halving

Explain that multiplying by 5 is also easy. If children can multiply by 10, and can halve numbers, multiplying by 5 is straightforward. Write down part of the 5 and the 10 times tables and show how the answers in the one are all half of those in the other.

$5 \times 1 = 5$	$10 \times 1 = 10$
$5 \times 2 = 10$	$10 \times 2 = 20$
$5 \times 3 = 15$	$10 \times 3 = 30$
$5 \times 4 = 20$	$10 \times 4 = 40$
$5 \times 5 = 25$	$10 \times 5 = 50$

multiplying by 3 by doubling, and then adding on the number

Explain that doubling can also be used to multiply by 3. You double the number and then add the double on to the original number.

$24 \times 3 = (24 \times 2) + 24 = 48 + 24 = 72$

additional strategies

partitioning into tens and ones

(see challenges 2, 5, 9, 10, 15, 16, 17, 18 and 20)

Multiplication Bingo

setting out the challenge

calculations
multiplying and dividing by 2, 3, 4, 5 and 10

getting started

For Year 3 multiply the numbers 2, 5, and 10 by the numbers 5, 10, 20, 25, 40 and 50. Write the results on the board. (For Year 4 multiply the numbers 3, 4, 5, and 10 by the numbers 5, 10, 20, 25, 40 and 50.

Ask the children to draw a 2 ¥ 3 grid (their 'Bingo card') and then to choose six numbers from the board to write on their card – one number per box.

Year 3	
10	100
20	125
25	200
40	250
50	400
80	500

a Bingo game for the whole class, then groups

class

Spin the two spinners and call out the numbers. Ask the children to multiply one by the other. Work with the children on calculating the answer. Discuss methods they can use to do these multiplications.

A child who has the answer to a multiplication on their grid should cross it out. The game is over when one child has crossed out all six numbers (or when everybody has crossed out at least one number).

Year 4	
15	100
20	120
25	125
30	150
40	160
50	200
60	250
75	400
80	500

Year 3
dice or spinner showing 2, 5, 10

dice or spinner showing 5, 10, 20, 25, 40, 50

Year 4
dice or spinner showing 3, 4, 5, 10

dice or spinner showing 5, 10, 20, 25, 40, 50

groups

Divide the class up into groups of four or five. Give each group a pair of spinners.

Each child selects six new numbers (from the same list on the board) to put on their bingo card. The group plays the game as before.

concluding

Choose sample multiplications to review the different strategies for multiplication used by the children.

"Can you do 50 times 3 by doubling and adding on?"

"Can you multiply by 5 using halving?"

variations

- Use spinners with more difficult numbers such as 4, 5, 9, 11 and 15, 25, 45, 50, 75, 100.

- Children write calculations (say, 3×5) on their Bingo card (this can be any number from 1 to 5 multiplied by 2, 5 or 10) and match the calculation against the product as the 'caller' calls the numbers.

Nine eleven

unpacking the strategies

mental strategies

multiplying by 9 by multiplying by 10 and adjusting

multiplying by 11 by multiplying by 10 and adjusting

expectations

Year 3: use patterns of similar calculations

Year 4: use closely related facts to carry out multiplication

Introduce these strategies during the challenge as appropriate.

multiplying by 9 by multiplying by 10 and adjusting

Explain the principle that multiplying by a number close to 10 can be tackled by multiplying by 10 and adjusting the answer.

$23 \times 9 = (23 \times 10) - 23 = 230 - 23 = 207$

Demonstrate this in the following way. Invite ten children to stand at the front of the class and each hold up, say, five fingers. Ask for agreement that ten lots of five is 50 and write this on the board:

$5 \times 10 = 50$

Now ask one of the ten children to sit down. There are only nine lots of five fingers left, because one lot of five has been removed. Check that there are 45 fingers now and write this on the board:

$5 \times 9 = 45$

What happened was that one lot of five fingers was removed.

$5 \times 9 = (5 \times 10) - 5$

multiplying by 11 by multiplying by 10 and adjusting

To demonstrate this strategy, invite another child to join the line of ten, and ask them to hold up five fingers. Clearly now there are eleven children, so, to multiply by 11 you can multiply by 10 and then add on another of the number you are dealing with.

$5 \times 11 = (5 \times 10) + 5 = 50 + 5 = 55$

and

$23 \times 11 = (23 \times 10) + 23 = 230 + 23 = 253$

additional strategies

multiplying by 10 and 100 by shifting the digits to the left **(see challenges 13 and 19)**

Nine eleven
setting out the challenge

a multiplication
challenge for
pairs and the
whole class

getting started

Write these numbers and operations up on the board.

15 16 17 18 19
20 21 22 23 24 25
  x 9   x 11

class

Explain that the 'Nine eleven' challenge is to get as close to 200 as possible using one of these numbers and one of the operations. (For Year 3 you might want to use smaller numbers and 100 as a target.)

target 200

Give the class a few minutes to work on one or two multiplications, individually or in pairs.

Then ask for some of their results. Talk about the methods the children used to multiply by 9 and 11.

Establish with the class which multiplication gets closest to 200. Both 18×11 and 22×9 make 198 (the closest answer).

$18 \times 11 = (9 \times 2) \times 11 = 198$
$9 \times 22 = 9 \times (2 \times 11) = 198$

Talk about why both multiplications give the same answer.

pairs

The children work in pairs and choose a new target between 100 and 300 (or zero and 100 for Year 3). They use the same numbers and operations as before to get as close to their new target as they can.

If time allows, the children play 'Nine eleven' three times, scoring a point for whoever gets closest to the target in the shortest time. The best out of three wins.

concluding

Ask the children to describe their calculations based on their new targets and review the different strategies used.

"If 5 tens are 50, are 9 tens more or less than 50?"

"Which is closer to 500, 9 fifties or 11 fifties?"

variations

■ A game for two players: each player in turn rolls two 1–12 dice and adds the dice numbers, then multiplies the answer by 9 or 11 to get as close as possible to 200. Whoever is closest wins a point.

■ Use one 1–6 dice and play the same game, aiming for 40.

■ Use two 1–20 dice and aim for 300.

Boxed in

unpacking the strategies

Introduce these strategies during the challenge as appropriate.

mental strategies

multiplying by 11 by multiplying by 10 and adjusting

multiplying by 9 by multiplying by 10 and adjusting

expectations

Year 3: multiply a single digit by 9 or 11

Year 4: multiply some two-digit numbers by 9 or 11

multiplying by 11 by multiplying by 10 and adjusting

Use diagrams to show how to multiply by 11. Explain, as an example, how to multiply 7 by 11. Draw a tower of seven squares and show how multiplying this by 10 results in a rectangle with an area of 70.

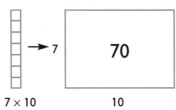

Go on to demonstrate how multiplying by 11 simply means adding on another tower of seven squares.

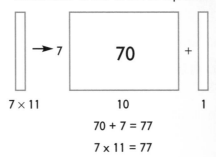

$$70 + 7 = 77$$
$$7 \times 11 = 77$$

multiplying by 9 by multiplying by 10 and adjusting

Use similar diagrams to show how multiplying by 9 involves a similar process to multiplying by 11 but by subtracting the original number, rather than adding it.

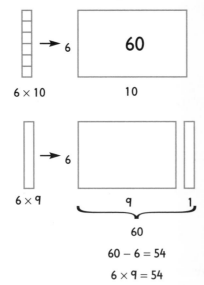

$$60 - 6 = 54$$
$$6 \times 9 = 54$$

additional strategies

multiplying by 10 and 100 by shifting
the digits to the left   **(see challenges 13 and 19)**

Boxed in
setting out the challenge

getting started

Start the lesson with some rapid calculation questions involving multiplying by 10.

Play the challenge 'Boxed in' against one of the children.

Part 1: Put a 4 × 4 square grid of dots on the board. Take it in turns to write a single digit (Year 3) or teen number (Year 4) in each of the 'squares' created by four dots (that is, nine numbers on a 4 × 4 grid). Numbers can be entered more than once.

Once the grid has been filled with numbers, take it in turns to join pairs of adjacent dots (vertically or horizontally; diagonally is not allowed). Whenever a player 'closes' a box (joins a pair of dots that completes the fourth side of a square), they claim the number in the square and draw a ring around it in their colour. Part 1 of the game is over when all the numbers have been ringed.

Part 2: Each player multiplies each of their claimed numbers by 11. They then find the total of all these products, using a calculator. They spin a coin: heads, and the player with the greater total wins; tails, the player with the smaller total wins.

a multiplication
game for pairs

different
coloured pens
for each pair

calculators
(optional)

```
3  5  1
2  5  6
3  1  8
```

```
3  5  1
2  5  6
3  1  8
```

③ ⑤ ①
② ⑤ ⑥
③ ① ⑧

Player One
3 × 11 = 33
2 × 11 = 22
3 × 11 = 33
5 × 11 = 55
8 × 11 = 88
Total: 231

Player Two
5 × 11 = 55
1 × 11 = 11
6 × 11 = 66
1 × 11 = 11
Total: 143

Tails – Player Two wins

pairs

The children play 'Boxed in' in pairs. Adapt the size of the grid (4 × 4, 5 × 5 or 6 × 6) and the choice of numbers to suit the age and level of attainment of the pair. When the children get to Part 2 of the game, note the strategies they use to multiply by 11 and, if necessary, teach them how to multiply by 10 and then adjust the answer.

If there is time, ask the children to play the game again, only this time to multiply all their claimed numbers by 9.

concluding

Ask the children to explain the strategies they used to multiply mentally by 11 or 9.

"If I claimed 5 and 7, what is my final score when I multiply each of them by 11?"

"Suppose I add the 5 and 7 together and then multiply by 11. Do I get a different total?"

variations

- Children write down four single-digit numbers and multiply each by 11. They then add up the single digits and multiply that by 11. Are the results the same?

- Spin the coin first to see if the winner is to be the one who has the greater or the smaller total.

Mental strategies

the full set

Use photocopies of this chart to assess progress.

class assessment

You can use the chart to record your observations and assessments while the session is in progress. Most of the challenges in this book involve children working for a while as individuals, or in pairs or small groups. This can give you an opportunity to note which strategies are being used, and which are not, for follow-up at the end of the session.

Highlight on a copy of the chart those strategies which you expect children to use, or which you have talked about in the first part of the session. Then walk around the class, or spend a while observing a table or group, and tick those which are actually being used. There is room at the bottom of the page to note down anything you may want to follow up – such as methods invented by the children themselves, or particular strategies that need further teaching. Use these notes to provide a focus for the wind-up to the session, or to help you plan future lessons.

individual and group assessment

You may also want to get a clearer picture of how individuals are working – to provide a baseline for comparison purposes in the future; to have a snapshot of their achievements at that particular time; to check whether they are working at full capacity; or to address concerns about any weak areas.

You can use the chart to assess up to six children over the space of a week or more. Write each child's name at the top of a column, and then use the middle part of the lesson to observe them, noting down those strategies which you see them using. Decide beforehand on a recording system that will be most helpful. For example, will you put a tick every time you see a child using a particular strategy? Will you identify those strategies that children appear to have difficulties with? How? Do you need to note the date every time you observe them, or will a more general statement, such as 'month/year', suffice?

a cumulative record

Some schools have a clearly defined strategy for observing the full range of children's potential, and keep detailed records, over time, of each child's achievement in specific key objectives. You can use the chart as a supplement to whatever recording sheet is used for numeracy as a whole – furnishing you with a list of the mental strategies in number calculation that children have demonstrated a knowledge of and prowess in.

strategies

putting the largest number first						
finding a small difference by counting up						
finding a large difference by counting back from the larger number						
counting on or back in ones, tens and hundreds						
partitioning into tens and ones						
identifying near doubles						
adding and subtracting a near multiple of 10, and adjusting						
using the relationship between addition and subtraction						
adding three small numbers by putting the largest number first						
adding three small numbers by finding pairs that total 9,10 or 11						
adding three two-digit multiples of 10						
using known number facts to add or subtract pairs of numbers						
using knowledge of place value to add or subtract pairs of numbers						
using doubling or halving, starting from known number facts						
adding or subtracting a pair of numbers by bridging through 10 or 100						
using patterns of similar calculations						
bridging through a multiple of 10						
doubling or halving two-digit numbers by doubling or halving the tens first						
multiplying by 4 by doubling and doubling again						
multiplying by 3 by doubling and then adding on the number						
multiplying by 5 by multiplying by 10 and then halving						
multiplying by 20 by multiplying by 10 and then doubling						
finding the 8 times table by doubling the 4 times table						
finding quarters by halving halves						
using closely related facts to multiply and divide						
multiplying by 9 by multiplying by 10 and adjusting						
multiplying by 11 by multiplying by 10 and adjusting						
using the relationship between multiplication and addition						
using the relationship between multiplication and division						
multiplying by 10 and 100 by shifting the digits to the left						
dividing by 10 and 100 by shifting the digits to the right						

Talking points in mathematics

Anita Straker

Cambridge University Press 1993

ideas that encourage children to use mental strategies

Issues in teaching numeracy in primary schools

ed. Ian Thompson

Open University Press 1999

essays by leading thinkers in primary mathematics education, including a practical emphasis

Teaching and learning early number

ed. Ian Thompson

Open University Press 1997

a review of research into the teaching and learning of early number concepts, including practical activities for teachers

Teaching primary maths

Mike Askew

Hodder & Stoughton 1998

activities, resources and case studies that make the link between the practical and the mental

The Numeracy File

Mike Askew and Sheila Ebbutt

BEAM Education 2000

articles on teaching primary mathematics – including mental strategies – with an emphasis on innovative and effective classroom practice

Teaching mental strategies

- *number calculations in Years 1 and 2*

Carole Skinner, Sheila Ebbutt and Fran Mosley

- *number calculations in Years 5 and 6*

Fran Mosley, Debbie Robinson and Mike Askew

BEAM Education 2001

the first and third books in the series

BEAM Education

BEAM Education is a specialist mathematics education publisher. As well as offering practical support for planning and teaching mathematics – from Nursery and Reception through to Key Stage 3 – BEAM actively contributes to research into how children learn mathematics and how best to teach it.

All BEAM materials are produced in close collaboration with mathematics specialists and teachers, and offer help with many of the current classroom concerns voiced by mathematics teachers. The materials are designed for use alongside other published schemes, to reinforce learning and stimulate discussion.

BEAM is dedicated to promoting the teaching and learning of mathematics as interesting, challenging and enjoyable.

BEAM services and materials include:

- courses and in-school training for teachers
- consultancy in mathematics education for government and other agencies
- a comprehensive range of teaching accessories
- over 70 publications for teachers of mathematics.